FLOWERS

of the Canadian Rockies

by Rick Langshaw

A Summerthought Publication
Box 1420, Banff, Alberta.

DEDICATION: To my brother Gary,
who loved plants and flowers.

ACKNOWLEDGEMENTS: I wish to thank
Rosemary Power for her
advice and help in proofreading.

ISBN 0-919934-15-3

Printed and bound in Canada

Edited by Thomas Owen

Published by
Summerthought Ltd.
P.O. Box 1420
Banff, Alberta T0L 0C0

Introduction

One of the most pleasant parts of any hiking or driving experience in the Rocky Mountains is the opportunity to see wildflowers. The mountains seem to enhance the natural beauty of multi-colored fields of alpine blossoms. Even on dark and gloomy days the flowers send forth an exuberance for life to all admirers. The mode of this admiration may take many forms: it may be photography, or a desire to recline amidst the lushness of an alpine meadow; the chance of seeing a rare simple flower; or a complex botanical study of relationships amongst plants. It is sometimes difficult for us as human beings to appreciate these relationships.

Perhaps you're one of the more unfortunate of our species who is relegated to a city—especially an office in the city. Before you succumb to the urge to rush into an alpine meadow to "feel the grass under your feet," or, to "pick just one flower for memory sake," try to consider the impairment you may be causing to a very delicate relationship. There are laws which attempt to protect the parks and these delicate environments, and it is the spirit of these laws which we must consider. Do certain things like flowers need protecting? Haven't they been looking after themselves for millions of years? The answer to this is not simple. It contains within its parameters the fact that there are now billions of human beings; that there are now problems like acid rain which have never existed before, that there are now introduced exotic species from other countries which out-compete the natural flowers, and so forth. So let us remind ourselves of these laws and why they are required. Their purpose is **PRESERVATION**, for those things which cannot protect themselves.

When writing a book on wildflowers the most important point to be considered is to give to the reader an enjoyment (through both the text and, especially, through the photography) of that which the author finds most pleasing about his subject. Beauty—the simplicity of this beauty—is the object of this book. All the illustrations are full color but, more importantly, the flowers presented are described in a non-scientific, easily understood manner. The method chosen for identifying flowers is by referring to the section which coincides with the color of the flower. For example, if the reader is looking for a blossom which is yellow, he or she should examine the section describing yellow flowers. The areas or regions where the flower is generally found growing (such as 'alpine' or 'sub-alpine') are defined in the case of each flower. Terms which are not common or in everyday usage have been explained in the glossary.

This book cannot offer the pleasures which are to be had by hiking to an alpine meadow, but hopefully it will serve to make this experience more pleasant once the work has been done to reach these truly exotic environments.

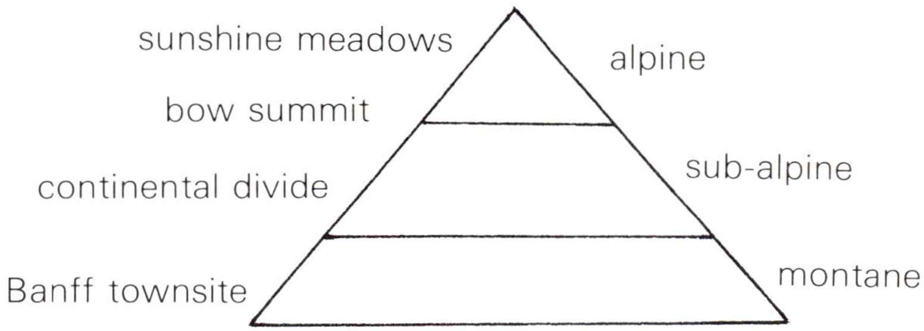
sunshine meadows
alpine
bow summit
sub-alpine
continental divide
montane
Banff townsite

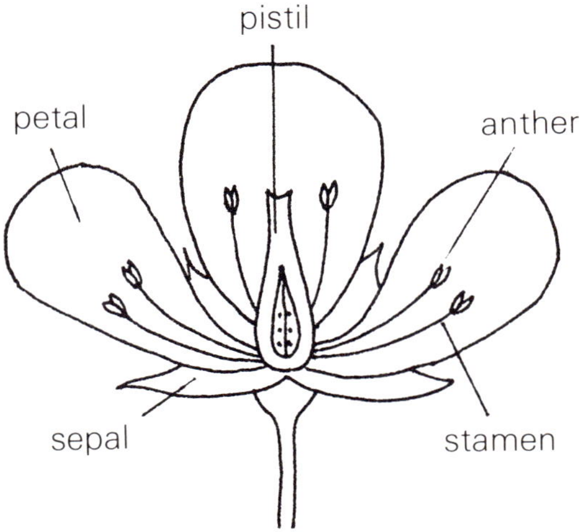
pistil
petal
anther
sepal
stamen

Arnica

There are about a dozen members of this family that inhabit this area. They are perennial plants which begin to flower in mid-summer and continue to do so until the autumn frosts. The leaves of this plant when crushed exhude a fragrance somewhat akin to expensive perfume. The leaves are heart or arrow shaped, arranged on the stem opposite to each other. All members of this family have bright yellow flowers.

Avalanche or Glacier Lily

This bright yellow lily is one of the first plants to bring color to the receding snowlines of The Rockies. As winter retreats, this flower at times grows right through the yet to melt snow. It is a true member of the Lily family, having a corm or bulb just under the surface of the ground. Rodents, bears and the deer family actively seek these corms as a source of food.

A noteable variation in this plant is the tendency for the stamens to vary in color from a chocolate red to a near white. This color variation may be noticed in just about any patch of lilies. From studies done on this peculiarity, it appears that the chocolate red is the normal color with other being a variant.

Bracted Honeysuckle or Twinberry

Bracted Honeysuckle is most attractive when it bears fruit. It's inedible purple-black berries contrast with the rich green, thickly veined leaves. The berries grow in pairs from a reddish cup-shaped arrangement of petals or bracts. The flowers are yellow, inconspicuous and also come in pairs. The bush is generally low to the ground and prefers damp, thickly covered montane or sub-alpine environments.

Arnica

Avalanche or Glacier Lily

Bracted Honeysuckle or Twinberry

Brown-eyed Susan or Gaillardia

This is one of the showier wild flowers of the Rockies. It is seen infrequently in most of the mountain parks, as it prefers sunny warm meadows in the montane grasslands. The east slopes of The Rockies, the vicinity of Radium Hot Springs, the western end of Yoho Park, and the low altitude prairie areas in Jasper are the most likely locations to find this exhuberant flower.

The flower grows well in domestic gardens and may out-compete its neighbors for grandeur.

One of the more uncommon properties of Gaillardia is its exotic, perfume-like scent. It is often possible to find the blossoms merely by sniffing the air and following one's nose. It flowers from mid-summer until early fall.

Canadian Goldenrod

This is a very wide spread plant in Canada and may consist of a hundred species, many of which are poorly defined. There are a number of fallacies concerning this plant; one is that it is a main contributor to hay fever (which has been disproven), and also that it has wide ranging curative properties (something which research has not been able to prove).

The stems may grow up to 2 meters in height atop of which is a spray of countless small golden flowers. This species should be differentiated from the Rocky Mountain Goldenrod, also found here, the height of which is seldom more than 20 or 30 centimeters, the flowers of which grow much larger but lesser in numbers.

Canadian Goldenrod is widespread in the parks through the montane region. It flowers in the fall.

Oregon Grape or Barberry

Upon first encountering this plant it appears to be Holly. The leaves have sharp pointed tips and have a habit of turning red in the late summer and autumn. Flowers come in numerous bunches and are bright yellow in color. The fruit is a dark blue and tends to hang beneath the leaves. This causes some difficulty in finding the berries at harvest time. The fruit is tart and acidic in flavour but makes fine jams and jellies.

Specimens of this plant are rare in Banff National Park but it occurs widely in Yoho, Kootenay, Waterton and Jasper Parks. Look for it on sunny dry hillsides in the montane region.

Brown-Eyed Susan or Gaillardia

Canadian Goldenrod

Oregon Grape or Barberry

Pineapple Weed

The smell of pineapple is the immediately recognizable feature of this "weed". This plant hugs the ground on most any area not readily usable by other plants. Pineapple Weed is a member of the Chamomile family and is often attributed with the same medicinal properties. It is an annual with numerous, smooth, fern-like leaves. The flowers are conical and composed of small yellow florets. The montane region is the upper limit of this plant's growth.

Senecio, Groundsel or Ragwort

The Prairie Groundsel and The Senecio triangularis are 2 members of a very large complex species of which there may be a score or more members in this area. They are mostly perennial herbs, about half of which may be found in the alpine region. At times they are quite difficult to distinguish. The Prairie Groundsel is up to 4 dm tall with leaves grey-green and fury underneath. The flowers have several heads with petals generally fuller and larger than those of the Senecio Triangularis. Look for fresh green leaves and pointed, arrow shaped leaves in the Senecio triangularis. It may be up to 1.5 m tall. Both of these may be found in the lower to mid-alpine, the latter especially in moist areas.

Shrubby Cinquefoil

This very showy bush is easily recognized by its bright yellow flowers usually flowering in mid-summer. Its name is derived from the French, noting that the flower has five petals. It is a deciduous plant which typically forms low rounded bushes growing up to 1.5 m high. It is common in all areas in the Rockies, especially in the lower alpine regions. In this century it has made inroads into all those areas which have been disturbed by man and horses.

Domestic, hybrid conquefoils have also been bred for garden use.

Pineapple Weed

Senecio, Groundsel or Ragwort

Shrubby Cinquefoil

Smooth Hawksbeard

This is one of the easily mistaken members of the Compositae family, most of which are hard to identify. It is an introduced annual which, like the dandelion, seems to be the scourge of open spaces, waste areas and gardens. It is, however, a very attractive plant, growing to nearly a meter in height and adding color to those areas which often lack it.

Hawksbeard is found throughout the parks in the montane region, flowering in late summer and fall.

Soap Berry or Buffalo Berry

This plant has been included under wildflowers not so much for its flowers (small yellow clusters of blossoms close to the leaf base) as for the plant's showy display of bright red, orange or yellow berries in the fall. When not in blossom this bush can be identified by the leaves which are leathery in texture, dark green in color, and furry on the underside. The fruit has a very undesirable soapy taste. The North American Indians seemed to have acquired a taste for these berries as they made a foamy sort of confection from them, often called "Indian Ice Cream". In the fall many mammals, most notable grizzly bears, seek these bushes for a particular type of feasting.

Tall Buttercup

The latin name of this common buttercup, (*Ranunculus Acris*) refers to its acrid taste. It is an introduced species which is quickly invading fields, pastures and waste areas. Domestic cattle eating this species have their milk rendered useless by its acrid taste. There have been numerous fatality reports of range animals eating this plant. If tasting this plant, do so with caution.

It is a tall plant growing up to 1 meter. In general, recognition of specific family members of the Buttercup is difficult.

Smooth Hawksbeard

Soap Berry or Buffalo Berry

Tall Buttercup

Western Columbine

This is one of the most appreciated plants of the Rocky Mountains. It has a wide distribution through the montane and sub-alpine growing regions. In these regions look for the Western Columbine in grassy meadows, in moist areas amongst boulders or on the fringes of the forest.

On the western slopes of the Rockies, Columbine species sometimes interbreed to produce colour variations (as depicted). The most common form is a pale yellow blossom with flared petals, which may turn slightly pink with age.

Butterflies and humming birds find this plant to be a great attraction.

Yellow Hedysarum

This is a member of the pea family. A close relative, Pink Hedysarum, is dealt with under Purple flowering plants.

This plant occurs in quantity on well watered avalanche slopes and moist south-facing slopes throughout the national parks. It grows in thick clumps seldom more than .5 m tall. The smell is sweet and the legume root is sugar-like in flavour. This factor seems to be the reason why both grizzly and black bears spend considerable time digging for the roots. Whole hillsides may show these scars. Yellow Hedysarum is most common in the upper montane and sub-alpine regions but may extend into the alpine.

Yellow Lady's Slipper

There are 2 members of this family common in The Rockies; the second, or "Franklin's Lady Slipper," is dealt with under white flowers.

The Yellow Lady's Slipper was once common in the Banff area but is now almost extinct due to OVER PICKING by people eager to collect one of these very showy orchids. Specimens are still to be found in Yoho, Kootenay and Jasper National Parks.

The plant blossoms very early in the summer and usually all flowering is complete within a few weeks. It appears the people who know where these orchids are abundant hide this fact to protect the plant. They are often found at about 1300 m in partially sunny, grassy clearings.

Western Columbine

Yellow Hedysarum

Yellow Lady's Slipper

Yellow Mountain Avens or Dryas

It is very easy to overlook this flower while it is in bloom. It is however a different story when this plant goes to seed. While flowering it is recognized as a pale yellow, nodding flower that does not fully open. Seeding proceeds first by appearance of a unicorn-like twisted spike of silk which then expands into a showy ball of golden-yellow. With the right reflection in the sun, this plant is a photographer's delight.

Look for Avens in gravelly floodplains in which it can expand for acres, unhampered by other growth.

Yellow Mountain Saxifrage

The Saxifrage family includes so many flowers that if upon trying to identify a particular flower of which you are not certain, it is a safe bet to say it is a saxifrage, and let someone else prove you wrong. The family has such a diverse assemblage and appearance that it is difficult even to give rough characteristics of what to look for when identifying this family.

Yellow Mountain Saxifrage is generally found spread like a carpet over stones near mountain streams. The leaves may be puffed with stored fluid, giving a succulent-like appearance. The stems are 5 to 10 cm long and generally bear one pale-yellow flower. The flowers are tiny, not more than ½ + cm in width. There should be 5 stamens.

Look for this flower in the alpine region, and also be alert for color variations (i.e. red dots).

Yellow Violet

Another member of this genus, The Western Canada Violet, is dealt with under white flowers. The yellow violet is quite common in the spring in moist, shaded and mossy areas. At times it grows in great numbers, with all the flowering usually completed within a few weeks. Look for it in the montane and sub-alpine regions.

Yellow Mountain Avens or Dryas

Yellow Mountain Saxifrage

Yellow Violet

Artemisia or Wormwood

Without doubt this is the most aromatic of alpine plants. A strong sage-like smell will alert the alpine hiker to the presence of this hardy plant. The leaves are reminisent of a fern and tend to be of a silvery hue. It is not a tall growing shrub (which is typical of most alpine plants), attaining perhaps 20 cm in height, most of which clings to the ground. It prefers calcareous soil on dry north-western exposures. If Wormwood is accidentally grabbed while scrambling on steep alpine slopes, the smell will linger very pleasingly on the hands for the rest of the day. The flowers are quite obscure, and in order to adequately examine the blossoms, a magnifying glass is required. Several similar species occur in all the mountain parks.

Cow-Parsnip

The latin name for this plant means "herculean," and indeed this plant is. It can grow to tremendous sizes, often to 3 m and more. Cow Parsnip roots are eagerly sought out by most of our wild herbivores and omnivores (grizzly bears especially). It is a perennial which grows heartily on avalanche slopes and in areas where there is abundant moisture. The blossoms are composed of dozens of smaller flowers arranged in a plate-like structure or umbrella. It extends through the montane and sub-alpine regions.

Death Camas

As the name would suggest of this plant, it is a truly poisonous species. Death Camas, along with Green or False Hellebore are two of the very few mountain plants which can cause death to the consumer. It is a perennial with a bulberous root. At the end of .5 m long stalks are groups of disagreeable smelling blossoms. Death Camas can be found in the montane and sub-alpine regions. Interestingly, the Blue Camas, a close relative, was an important food for native indians.

Artemisia or Wormword

Cow-Parsnip

Death Camas

Elderberry

Although this shrub may appear at times to be a tree, due to the large size it attains (5 m), it truly is a shrub. In the Rocky Mountains it is a resident of the montane and sub-alpine regions. The large, dark-green, saw-toothed leaves are usually arranged in 2-3 pairs of opposite leaflets. Elderberry is early to unfold its greenery in the spring and immediately sheds its leaves with the first frost, leaving behind clusters of blue-black berries. Likewise, the flowers are found in large white bunches of fragrant blossoms. The berries are famous for the jams, jellies and wines which can be produced from them.

High-Bush Cranberry

As this is not a very conspicuous plant, it is easily overlooked by most people. The berries have a very delicious flavour, especially if picked after the first autumn frost. The bright red berries remain on the bush all winter, and therefore are an important source of winter food for animals. The bush attains a height of about 2 m in the montane regions but may be considerably shorter as it approaches timberline. The place to find this delicious member of the Honeysuckle family is in damp woodlands and on thick, old avalanche paths. Caution is advised as these berries are similar to the inedible Twinberry.

Leafy Lousewort

This is a rather peculiar name for such a delicate alpine plant. There are several members in the Lousewort family. The Leafy Lousewort is identifiable by its remarkably twisted pale flowers. It is a perennial, which grows from 10 to 50 cm in height and is usually associated with conifers on high sub-alpine to alpine slopes. The flowers are greenish-white with traces of yellow or purple.

Elderberry

High-Bush Cranberry

Leafy Lousewort

Milfoil or Yarrow

Yarrow is an old Scottish name, after the parish Yarrow on the little river of the same name. This plant is an aromatic perennial with a smell very mush like Sage. The flowers are flat-topped clusters of small white to pink flowers. The clusters are on a strong wirey stock which may reach .5 m in height. This plant prefers dry, sunny montane regions and can be found flowering throughout the summer and fall. The leaves are fern like and finely divided. Herbalists ascribe many curative properties to this plant.

Ox-Eye Daisy

The Fleabane, Aster, and Daisy family are very confusing to most people interested in mountain flowers. These plants are all very closely related and it would serve no purpose here to try and differentiate between them. They all have composite blossoms and even within a particular species may have differing colors. The Ox-Eye Daisy is found in the montane, sub-alpine and alpine regions. It often grows in very large numbers and may appear from a distance to be snow on a mountain meadow. It is particularly common on roadsides.

Thimbleberry

The Thimbleberry, like its very close cousin the Wild Raspberry, produces white flowers which eventually become red, seedy fruit, eagerly collected by people, bird, rodent and other omnivores. The Thimbleberry is distinguishable from the Raspberry by its much larger leaves (up to 20 cm broad) and its preferred habitat (that of wooded sub-alpine moist areas). The Raspberry prefers sunny, warm, south facing, dry, montane and sub-alpine slopes. The fruit of the Thimbleberry is much seedier and less tasty than its cousin the Raspberry.

Milfoil or Yarrow

Ox-Eye Daisy

Thimbleberry

Valerian

This is a very interesting alpine plant for several reasons: it is very much like catnip in its attraction for domestic cats; it has been used as a sedative in herbal medicine for centuries, and, finally, in its smell; a sickeningly sour aroma after the first fall frost. The stalk is up to .8 m tall and is topped by a cluster of blossoms, white in color, usually hemispherical in shape. The growing region of Valerian is sub-alpine and alpine, where it prefers damp meadows.

Western Canada Violet

This is an attractively tall member of the Violet family found in the Rocky Mountains. It often grows up to 40 cm tall, when found in its preferred habitat—rich aspen forests. The flowers are larger than those of the Yellow Violet and the leaves are also larger and heart shaped.

White Bog Orchid

This flower is truly one of the treasures of the montane and sub-alpine swamps and bogs. The stalk can reach heights of 1 m, and it may hold up to 20 or more fragrant white blossoms. One of the chief identifying marks of this orchid is the spur-like lip on the flower, which is slender at the tip and quickly expands or dilates at the base. Look for the White Bog Orchid in sunny bogs and marshes during the early summer.

Valerian

Western Canada (Wood) Violet

White Bog Orchid

White-Flowered Mountain Avens

This plant, like other members of the Dryas family, forms large colonies in dry, stoney places where there is ample sun and little competition. Large mats of these freely rooting plants are found in all alpine and montane areas where the right conditions exist. The leaves are leathery with the underside coated with felt-like hair. This is for protection from the harsh climate which they inhabit.

The leaves tend to exude a spicy fragrance on warm sunny days. This aroma is difficult to pinpoint as to its source. If one or two leaves are rolled between the fingers, this exotic smell will stay on the hands for hours. The white blossoms evolve into a silky sail, which aids in the dispursement of the seeds.

White Globe-Flower

The White Globe-Flower is one of the earliest of the alpine plants to flower in the spring. These plants produce one or, rarely, two flowers. It prefers quite damp, marshy or stream-side environments. It is often found in conjunction with Alpine Buttercup and Marsh Marigold.

White Pasque Flower

Of all the alpine plants this certainly has the most impressive fruiting structure. The plain white flower which appears in early summer quickly becomes a very hairy ball of silk, perched atop a stalk about 30 cm in height. Entire alpine meadows become waving fields of these heads of uncombed hair. Its preferred habitat is in rich herb mats of the sub-alpine and open alpine meadows.

White-Flowered Mountain Avens

White Globe-Flower

White Pasque Flower

Dwarf Dogwood or Bunchberry

This plant must be considered as one of the very special plants of the mountain woods. Regardless of the time of year, this brave little subshrub is exuberant and showy. If uncovered during the winter, Dwarf Dogwood will still possess red berries and perhaps leathery red leaves. It is quick to "green-up" in the spring and will unfold small white blossoms early in the summer. These blossoms in turn produce a tight cluster of spring and will unfold small white blossoms early in the summer. These blossoms in turn produce a tight cluster of berries, usually recognized as "bunchberries," in the late summer and fall. By late fall the berries, along with the leaves, become a radiant crimson. At this time the entire forest floor may become a single carpet of dogwood-red.

Dwarf Dogwood will grow almost anywhere there is rotting wood, moisture and ample shade. It is very difficult to pinpoint a growing zone for this plant, as its distribution is so wide. It is suffice to say, however, that it can be found throughout The Rocky Mountains except in the alpine.

Dwarf Dogwood or Bunchberry

Mountain (White) Lady's Slipper

Like its closely related cousin the Yellow Lady's Slipper, this is a member of the orchid family. The stems are up to 25 cm high growing from a fibrous rootstock. The flowers are usually solitary but may be occasionally double or even triple. The blossom has a fine fragrance and is seemingly over-endowed with an exquisite white pouch, backed with brown sepals and petals. Its usual habitat is mountain slopes in rich moist humus. Occurring at moderate altitudes, it is always associated with tree cover in close proximity. Of all our mountain flowers, this is one of the most endangered by transplanters and pickers. Any disruption of this plant will kill it, so only observe this orchid—please do not touch it!

Twisted Stalk

It is simple to confuse Twisted Stalk with plants like the Fairy Bell. The distinguishing feature is that each flower hangs separately from a uniquily kinked stalk. The berries are bright crimson and should not be eaten by humans. In the Rocky Mountains this plant is most commonly found in damp, forested, montane and sub-alpine regions.

Mountain (White) Lady's Slipper

Twisted Stalk

Colt's Foot

This plant is one of the earliest to push leaves through the ground in the spring, often while snow is still on the ground. During the growing season the leaves may reach 30 cm in width. It is often difficult to associate the leaves with the blossoms of this plant, as the flowers may arise from a creeping rhizome some distance from the leaf. Upon maturing, the flowers produce a dandelion-like puff, which aids in dispurtion of the seeds. Colt's Foot may be found in the montane and sub-alpine. A damp, shaded, herb mat is the preferred locality.

Labrador Tea

This is one of a small number of evergreen shrubs in the mountain parks. The leaves easily identify the plant. They are dotted with glands containing resin, are dark glossy-green on top and have a rust-colored, wooly, lower surface. The leaves contain Ledol, a toxic substance which can cause poisoning even when ingested in small amounts.

Numerous white flowers appear from May through July on the tops of the stems.

Historically, Labrador Tea served as a tea for native indians and was also used as a smoking tobacco by the Coureurs de Bois.

Colt's Foot

Labrador Tea

False Dandelion

This perennial member of the *Asteraceae* family grows in the sub-alpine to alpine regions. Generally, it can be found in meadows or areas with plenty of sunshine. Its reddish-orange blossom is about the size of a 50 cent piece, with the stalk growing to approximately .5 m in height. False Dandelion is not a common plant and it may be difficult to find a specimen if looking for one.

Indian Paintbrush

As should be apparent by the photograph, these showy members of the CASTILLEJA family can be found in a great many colors, ranging from bright red through orange to yellow or green. Each color variation is a sub-member of the family and may show some varying preference to habitat. The entire family however is sub-alpine to alpine in region and generally prefers south-facing, well-drained meadows. An interesting aspect of this plant is that it is a root parasite, requiring certain other plants in close proximity as hosts.

False Dandelion

Indian Pairtbrush

Orange Hawkweed

This is a very attractive, bright orange flower introduced from Europe. Orange Hawkweed spreads rapidly by winged seeds and also creeping rhizomes or roots. Introduction of this plant into the mountain parks was a mistake, as Orange Hawkweed takes over very quickly in these environments and out-competes most other native plants. Look for this flower along roadsides and waste areas in the montane region.

Western Wood Lily

The Western Wood Lily is one of the most visible members of the Lily family. The vivid orange-red coloration of the blossom, and a stock of up to .6 m in height, immediately sets this flower apart from the habitat in which it lives. Look for this plant in montane grassland meadows and sub-alpine meadows. It requires a moist soil condition and plenty of sun. The Western Wood Lily used to be common throughout the Rocky Mountain Parks but in recent years has suffered seriously from being picked.

Orange Hawkweed

Western Wood Lily

False Azalea

The flower of this plant tends to be non-descript; however, this 2 m high bush becomes quite attractive in the fall when the leaves turn orange and crimson. During the summer the light green leaves easily identify the False Azalea amongst the other bushes and shrubs in which it lives. Its preferred habitat is moist mountain stream banks and shaded woods in the montane and sub-alpine regions.

Red Clover

Here is a very familiar plant introduced to roadsides, waste areas and cultivated fields. A member of the Pea Family, it has a nitrogen fixing root system which is held in high regard by farmers for fallow fields. The taste of the flower is very sweet, for they contain a good amount of the raw material for honey making. A light green to white crescent should be noticeable in the centres of the leaves. Red Clover plant is restricted to the montane regions of all the mountain parks.

Twin-Flower

The family name of this exquisite little flower is Caprifoliaceae. The name comes from the zodiacal sign of Capricorn since the members of this family have twisted stems for climbing, like a goat's twisted horns.

Twin-flower spreads runners out from the main plant, which creep over moss and dead foliage with stalks sprouting at 5 to 10 cm intervals. Each stalk supports a pair of trumpet-like, pink flowers. The fragrance emitted by these blossoms is sweet and spicy.

The Latin name of the plant is Linnaea borealis, named after Carolus Linnaeus, the famous botanist.

The montane growing region is the preferred habitat of this species, where it likes cool, shaded places.

False Azalea

Red Clover

Twin-Flower

Lyall's Larch

It would seem strange to include a tree in a book on wildflowers were it not for the exotic-looking blossom which this tree produces. This member of the larch family is sub-alpine to alpine in its growing region and often is the only forest cover near timberline in the southern parts of Banff and Yoho Parks. It is widespread in Kootenay National Park but very rare in Jasper Park. One of the most notable factors concerning this coniferous tree is that its needles turn bright yellow and fall off about mid-September, (unlike other conifers).

Alpine areas such as Larch Valley in the Lake Louise district of Banff Park, the Lake O'hara area of Yoho Park, and the Wolverine Pass area in Kootenay Park become a blaze of gold as the larches turn yellow.

Mountain Sorrel

This interesting flower derives its technical name from the fact that it has a sour flavour when eaten. It is very common in the alpine growing regions and has been found in The Rocky Mountains at heights upwards of 3000 m. The leaves are rounded and noticeably veined. The flowers are found atop stiff, long stems and are reddish-brown in color. Hikers in the Rockies can be rewarded with a lemon-like, sharp flavour if lucky enough to taste a leaf of Mountain Sorrel during their travels. It is wise not to eat more than just a taste, as the acid which produces this flavour is poisonous in large amounts. Look for Mountain Sorrel in wind-protected, damp soil.

Lyall's Larch

Mountain Sorrel

Kinnikinnick or Bearberry

There are many tales about the use that native Indians have had for this plant. Most stories deal with the use of this plant for smoking in a pipe. It seems that when mixed in the right proportions with tobacco this does provide a passable smoke. Kinnikinnick grows very low to the ground, forming mats as it spreads in all directions. The branches are flat with reddish, shreddy bark and leathery, evergreen leaves. The blossoms are pink with a gradual reddening as the end of the flowering is reached. The berries are mealy and not pleasant to eat. They form an important source of food for bear. It is common on dry, south-facing hillsides in open coniferous woods.

Purple Aster

This is a member of a large, hybridizing genus represented by about 2 dozen members in the national parks. The Asters are closely related to the Fleabanes and, at times, it is hard to distinguish the two. Both Aster and Fleabane flowers have yellow or purplish rays and yellow (or occasionally slightly purplish) centres. Fleabanes may be distinquished from Asters by their thinner, more abundant rays. Asters usually have less than 50 rays per flower, Fleabanes usually have more. The Purple Aster is one of the sub-alpine to alpine region members. It is usually found on sunny and dry, grassy slopes, flowering in late summer to fall, and is often one of the year's last blossoms.

Purple Aster Kinnikinnick or Bearberry

Wild Onion

Suprisingly enough, this is a member of the Lily family. It is not hard to recognize; it has a very strong smell of onion. However, it is wise to remember that there are a great number of plants in this family, several of which are deadly poison. The leaves are flat and fleshy and are invariably slightly shorter than the stem which bears the flower. Look for the Wild Onion in wet sub-alpine meadows. It flowers in early to late summer and leaves a characteristic seed-head. The leaves will provide a very pleasant addition to a backpackers supper, but it is best not to dig the bulb as this destroys the entire plant.

Wild Rose

The Wild Rose was well chosen to be the emblem of the Province of Alberta. It is probably one of the most prolific of flowers of roadsides in the montane and sub-alpine regions. The bushes may reach 1.5 m in height and at times will be covered in reddish-pink to almost white flowers.

The fragrance of this member of the rose family is unsurpassed and the blossoms are often collected for just this reason. Near the end of August the fruit of the blossom, sometimes called "rose hips," begins to ripen and turn red. For herbalists and jelly makers alike these berries are a real treasure. The Wild Rose prefers dry south-facing slopes where the summer temperature becomes quite hot. This plant is rarely found above timberline.

Wild Onion

Wild Rose

Apetalous (Bladder) Campion

Upon first viewing this peculiar little flower, the impression is that one is looking at a chinese lantern. The blossom is unique in this respect and therefore the key factor in its identification. The Apetalous Campion is restricted to the alpine region on sliding rock slopes. The flower is slightly pink or purplish and in most specimens does not open. When open, small pink petals near the mouth of the globe-like flower are seen. It is to be found flowering in mid-summer.

Bog Epilobium

This small and dainty flower is a close relative of the Fireweed, being of the Evening Primrose family. It is found in the sub-alpine to alpine regions in damp lush areas by streams or lakes. Its most easily identified structure is the tube-like seed pod which may grow to 5 cm long. The pod is present even before the flower blooms. At this time it looks like a slightly swollen, pinkish flower stem. The pod is typical of all members of the Fireweed or Epilobium genus. This pod splits with age, releasing seeds covered with tufts of silky hair which easily becomes airborne. As with the Fireweed, it flowers in mid to late summer.

Bull Thistle

This is an introduced species which unfortunately does very well throughout most of North America. Commonly, it is found along roadsides, waste areas or in unkept gardens. It is a very formidable plant being well armoured with hundreds of sharp prickles against all oncomers. The purple flowers are about 2 to 4 cm in width and exude a sickly sweet aroma. The Thistle can grow to 1.4 m in height, which presents a truly impregnable wall for those intent on its undoing. For the intrepid, the roots and peeled stems will make a passable emergency food.

Apetalous (Bladder) Campion

Bog Epilobium

Bull Thistle

Elephant's Head

This peculiarly shaped flower is a member of the Lousewort family. It is immediately distinguishable from other members of its family by its very long (up to 25 cm) stalk of purple-red flowers. When looked at closely, the individual flowers appear very much like the head of an elephant. This resemblance is accentuated by the "upturned trunk" portion of the flower.

Elephant's Head is found in rich, damp, subalpine meadows. It prefers calcareous soil, particularly with mineral springs which offer it both plenty of moisture and correct mineral supply.

Fireweed

Here is the great colonizer of disturbed areas and recently burned forests. Shortly after a major disruption to an area, like a fire, this plant moves in in tremendous numbers. Vast expanses of valley bottom and hillsides will become swaying carpets of pink, as the fireweed flowers in late summer and fall. The seeds are mini-parachutes of gossamer, filling the air with the slightest breeze. As frost moves into the valleys in the fall, the Fireweed plant turns to deep colors of orange and flaming red. Be it for the color, or for the fact that it moves in after a fire, one thing is for certain — the Fireweed plant is aptly named. It is a resident of the montane, sub-alpine and alpine regions.

Fleabane

With about two dozen members, it is a very large family of difficult-to-distinguish perennial plants. More than half of the genus are residents of the alpine zone, the others are scattered through montane and sub-alpine regions. In order to distinguish the Fleabane from its close relative, the Aster, it should be remembered that Fleabanes flower in spring and early summer, while Asters bloom in the late summer. Fleabane blossoms may be yellow or purple through to pink or even white. Yellow centres to the blossoms are the rule. The name Fleabane comes from an old belief that the dried plant would deter fleas.

Elephant's Head

Fireweed

Fleabane

Heal-All or Self Heal

The name of this plant bears witness to the reputation the plant once had in its ability to cure various types of ailments. No certain cures have come from various extracts of this plant, but it is still highly regarded by herbologists. It is found growing in the montane through to the sub-alpine regions. The plant prefers cool shady environments where moisture is available regularly. The blossoms are usually deep purple but on occasion may vary to pink or white. Heal-All is often found in groups, with group members growing from one original parent-plant.

Horse Mint or Wild Bergamont

Here is a typical member of the mint family. It is easily recognizable by its leaves, which are sharply toothed and opposite on the stems, and by its aroma which is menthol or, to some, like a Bergamotte Orange.

This plant is not common in the Rocky Mountain Parks. It can be found in Waterton Park and on the east slopes of Banff National Park in low montane grasslands. It is uncommon in the other mountain parks.

The flowers vary in color from blue through to pink and near white. The blossoms always occur in bunches at the top of the stem.

Moss Campion

This flower is truly one of the most representative of alpine species. It is found almost exclusively in rocky areas in the high alpine where it forms tight cushions, which may be slightly domed and reach widths of 1 m or more. Moss Campion can be found flowering just shortly after the winter's snow has receeded and may continue to flower into midsummer. The pink flowers are large in comparison to the rest of the plant. The leaves are tiny and needle-like, growing low to the ground to escape the wind and prevent moisture loss.

Heal-All or Self Heal

Horse Mint or Wild Bergamont

Moss Campion

Mountain Heather or Heath

The Purple, Yellow and the White Mountain Heathers are closely related. They have a similar habitat and range. White Heather has small, scale-like leaves, Purple Heather has somewhat shorter leaves, which are grooved on both sides. Yellow Heather has leaves of intermediate length, lacking the groove. The Yellow and Purple species tend to form a hybrid when in close contact. The product is a pink flower. Look for all of these plants in the alpine zone on damp acidic soils.

Penstemon or Shrubby Beard-Tongue

This plant belongs to a very diverse family which are almost entirely restricted to North America. The family encompasses over two hundred members of which we have about a dozen in the Mountain Parks.

Shrubby Beard-Tongue is partially evergreen in the fact that each autumn only some of its leathery leaves turn red and fall, while others are left green for the winter. The blossoms can be quite large, growing upwards to 5 cm.

Look for this plant growing on rocky ground in the montane through to the alpine regions. It is not unusual to find specimens growing at 3000 m in altitude.

Pink Hedysarum

This is a member of the pea family. Yellow Hedysarum, a close relative, is dealt with under Yellow flowering plants.

Pink Heysarum occurs in abundance on well watered avalanche slopes and moist south-facing slopes throughout the national parks. It is seldom more than .5 m tall but can grow in thick bunches. The smell is sweet and the legume root is sweet in flavour and high in protein. This factor seems to be the reason why both grizzly and black bears spend considerable time digging for the roots. Whole hillsides may show their scar. Pink Hedysarum is found through the montane and sub-alpine regions.

Mountain Heather or Heath

Penstemon or Shrubby Beard-Tongue

Pink Hedysarum

Purple Hawkweed, Saussurea or Saw-Wort

This flower is not uncommon in the alpine growing region but, due to its habit of growing on very rocky slidepaths at high altitudes it is poorly known. The blossoms, usually a bright purple, intermixed with white wooly hairs are bunched at the top of the stem. The leaves have a soft down which protects them from the harsh climate in which they live.

Purple Saxifrage

This flower can easily be mistaken for Moss Campion so look for these identifying features: The leaves are wedge shaped, three but rarely five-lobed; flowers are one to three in groups, the petals most often being pink, turning to white as they get older. The stems, unlike Moss Campion, are not woody. Look for Purple Saxifrage in moist gravelly areas in the Alpine zone. Flowering is usually complete by August.

Purple Virgin's Bower or Clematis

This flower is a resident of woody glades and fallen timber areas, where it creeps and climbs in profusion. It is one of the few vines found in the mountains and therefore is readily recognizable. The blossoms are pale purple or blue. A white flowered species grows in Waterton Lakes National Park.

Domestication of this plant has occurred in many areas. It is now often found climbing along well manicured garden trellises.

Purple Hawkweed, Saussurea or Saw-Wort

Purple Saxifrage

Purple Virgin's Bower or Clematis

Shooting Star

This small, exquisitely flowered plant is easy to recognize by its backswept pink petals which reveal a yellow and purplish-black centre. The Shooting Star is found in sub-alpine meadows which receive generous amounts of sun and abundant moisture. The beauty of this flower has led to a serious decrease in its population due to indiscriminate picking. One related species, the Littorale Shooting Star, has already become extinct.

Venus's Slipper or Calypso Orchid

This is a true member of the orchid family. It grows from a corm or tuberous rootstock, which is all that remains after the plant flowers and seeds in early summer. Montane forest is the prefered habitat of this orchid, where it usually grows amidst Horsetails and Dwarf Dogwood. As with all orchids, this flower grows only in association with a certain type of fungus. As a result it is virtually impossible to transplant.

This plant suffers from over-picking by collectors. Picking of this flower will kill the plant, so it is best to remember to leave this as well as all flowers to mature unimpaired.

Shooting Star

Venus's Slipper or Calypso Orchid

Alpine Forget-Me-Not

A flower of early summer, it grows abundantly in dense clusters in the sub-alpine to alpine regions. The blossoms are very fragrant with petals a light blue, aging to pink with a bright yellow centre. As you increase in altitude, the flower becomes a much brighter blue. The stalks can grow to .5 m in height when the plant is found in association with other tall plants. Generally, however, it is 10 to 20 cm tall. The Forget-Me-Not is a perennial and adapts well to domestication in the towns of Jasper and Banff, but does not transplant well out of its native habitat.

Delphinium or Larkspur

This attractive annual is found throughout the Rocky Mountains in montane and sub-alpine growing regions. Depending upon altitude and the type of soil in which it grows, it can attain heights of between 10 cm and 1 m or more. Generally, in the sub-alpine, it is quite short. The Larkspur can cause serious problems for animals which graze upon it. A type of poisoning which manifests itself as intestinal cramps and diarrhea, and in some cases death, is common where this plant grows in pastures.

Color of the Larkspur is variable from white to shades of blue or violet. The blues and violets with tinges of white are the most common color variations.

Prairie Crocus or Pasque Flower

The name Crocus when applied to this plant is incorrect as it is a true member of the Anemone family. This is one of the first flowers of spring, usually found on open grassy slopes. The blue flowers appear earliest in the prairies, often as soon as April in areas close to Calgary. Generally, however, these blossoms are not seen in the mountainous areas until May or later. The Pasque Flower is rare in most areas of the Mountain Parks but can be found on the east slopes of the Rockies in Banff, in Waterton, and in the rolling grassland areas of Jasper.

Alpine Forget-me-not

Delphinium or Larkspur

Prairie Crocus or Pasque Flower

False (Green) Hellebore

This genus has nearly a dozen members in North America but thankfully it is represented in this area by only one, Green Hellebore. Thankfully, as this plant is deadly poison. It is one of a few in the mountains which is poisonous. Green Hellebore is a very attractive plant with large, full leaves which are strongly parallel-veined. The most impressive time to view this plant is in spring, just as it emerges from the ground. The emerging plant appears like the nosecone of an ascending rocket. It is at this time which it is most poisonous. The green blossoms are disappointing in their lack of exuberance, which one does not expect from such a lusty plant as this. Green Hellebore inhabits cold, damp alpine and sub-alpine meadows. With the first frost the leaves become bright yellow and thus a photographer's delight.

False (Green) Hellebore

Index

Glossary

Montane: Forests and grasslands found at low elevations such as the areas surrounding Banff, Jasper and Field townsites. Montane habitat stretches from the valley bottoms up to 1600 m.

Sub-alpine: An example of the sub-alpine is at the point where the trans-Canada highway crosses the continental divide near Field, B.C. Here at elevations of between 1600 and 2000 m the growing season is shorter and the winter harsher.

Alpine: This is the "true high country" of the Canadian Rockies. At elevations of over 2000 m trees are either not present or are stunted. The land may be carped with a rich meadow of flowers. Some of the most easily accessible alpine meadows are those in the Sunshine Village ski area or those just a short hike from the road crossing Bow Summit.

Acrid: bitter

Calcareous: soil which contains carbonate of lime or limestone, this soil is alkaline.

Conical: cone-shaped

Fallow: fields which have been plowed and left without planted crops, usually wild plants have seeded themselves in fallow fields.

Gossamer: a thin film or web

Rhizome: a horizontal, fleshy underground stem resembling a root

Sepal: the green or colored outer leaves or scales covering a flower bud; once the flower opens these sepals are usually seen at the base of the flowers

Stamen: the male or pollen bearing portion of the flower composed of a stalk or filament and a pollen head or anther.